The inside cover photograph shows a section through the West Runton Freshwater Bed about 1.5 metres deep. Laminated sands and silts beneath; coarse sandy silts with stones at base, dark peaty silts above. Note the abundant freshwater shells.

ISBN 0 903101 64 5

Design and production by Em En Designs, Norwich.

Printed by BD&H, Norwich.

Acknowledgements

The West Runton Elephant Project is supported by the Heritage Lottery Fund; MGC/Science Museum PRISM Grant Fund and Anglian Water.

Thanks are also due to Bernard Matthews plc; English Nature; Economic Development Unit, Norfolk County Council; Friends of the Norwich Museums; Arkeologikonsult, Sweden; David Abbs; Richard Abbs; Tarmac Roadstone; Brian Farrow, Peter Lawton and other North Norfolk District Council Technical Services staff; Renosteel; Fiona Beatty; Harold and Margaret Hems; Rob Sinclair; John Clayden; Kevin Johns; Jimmy Lightwing; David Blackman; Sonia Smith and Julie Curl; the many volunteers who assisted with security and other tasks during the excavation; Norfolk Museums Service staff: Trevor Ashwin and Sarah Bates, and staff of the Norfolk Archaeological Unit; Mary Davis, Dr Tony Irwin, Rob Driscoll, John Goldsmith, Brian McWilliams, Cathy Proudlove, Martin Warren, Nick Arber and the Display department, Dr Gordon Turner-Walker, Elena Makridou, Dave Harvey, Sarah Hollis, Susie West, Nigel Larkin, Stephen Drew and the Finance Department.

THE WEST RUNTON ELEPHANT
DISCOVERY AND EXCAVATION

Norfolk Museums Service

THE WEST RUNTON ELEPHANT

DISCOVERY AND EXCAVATION

by A J Stuart, BSc, PhD, DSc.

Travelling around Norfolk today, we see a gently rolling landscape, with fields, woodland, heaths, wetlands and built up areas. A varied North Sea coastline ranges from cliffs to salt-marshes and sandbanks. Although much wildlife is still to be found, nature has retreated everywhere in the face of thousands of years of human activity. This is hardly a part of the world that one would associate with elephants, rhinos or giant moose. Yet fossil finds tell us that these and many other animals once roamed the forests and grasslands of what is now East Anglia.

From time to time powered by north-westerly gales the sea pounds against the soft crumbling cliffs along the North Norfolk coast - exposing the layers of sediment laid down hundreds of thousands of years ago during the Quaternary period - known informally as the Ice Age. The continuing retreat of the cliffs also brings to light fossil teeth and bones, all that remains of the ancient inhabitants of the region.

This is the story of the discovery and excavation of the most spectacular of these fossil finds, the oldest and largest fossil elephant skeleton ever to have been found in Britain.

The Cromer Forest Bed

Most of Norfolk is underlain by Chalk, a kind of limestone deposited in a subtropical sea between about 70 and 97 million years ago (late Cretaceous period). On top of the chalk are extensive deposits of much younger sands, gravels, clays and other soft sediments of the Quaternary period, dating from about one and a half million years ago up to the present day.

The Quaternary period (Ice Age) was by no means uniformly cold, but showed marked changes in climate on a time scale of a few thousand years. In Britain deciduous forest clothed the landscape in the warmest phases, or interglacials, while treeless steppe (prairie) or tundra-like vegetation grew during in the cold phases. At times during these cold phases ice sheets covered extensive areas of northern Europe including Britain.

About 400,000 years ago, in the Anglian Cold period, Britain was in the grip of an arctic climate, and an ice sheet several kilometres thick extended over most of Britain, as far south as what is now north London. Boulder clay - debris comprising clay, sand, gravel and boulders - was dumped by the melting ice, while layers of sands and gravels were deposited by escaping meltwater. These glacial sediments blanket much of East Anglia and can be seen at many places on the coast exposed in the cliffs, where the sea has eroded the land.

Exposed in the cliffs and on the foreshore, beneath the boulder clay and outwash sands and gravels, and taking us further back in time, are very different sediments known as the Cromer Forest Bed Formation or 'Forest Bed' for short. It is exposed at intervals along the coast of Norfolk and adjacent Suffolk, from Weybourne to Kessingland. The Forest Bed (the name comes from occasional finds of fossil tree stumps) includes various kinds of sediments, some deposited in the sea and others in rivers. The older sediments, including the marine shelly sands known as 'Weybourne Crag', date from about one and a half million years ago, whereas the younger group of deposits was laid down from about 700,000 to 500,000 years ago. The climate was at times temperate, much as today, and at other times subarctic.

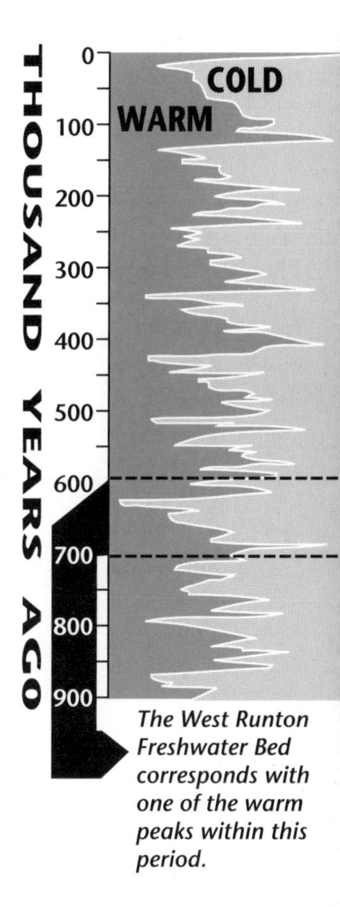

Changes in climate over the past 900,000 years.

Map of Norfolk showing Cromer Forest Bed locations

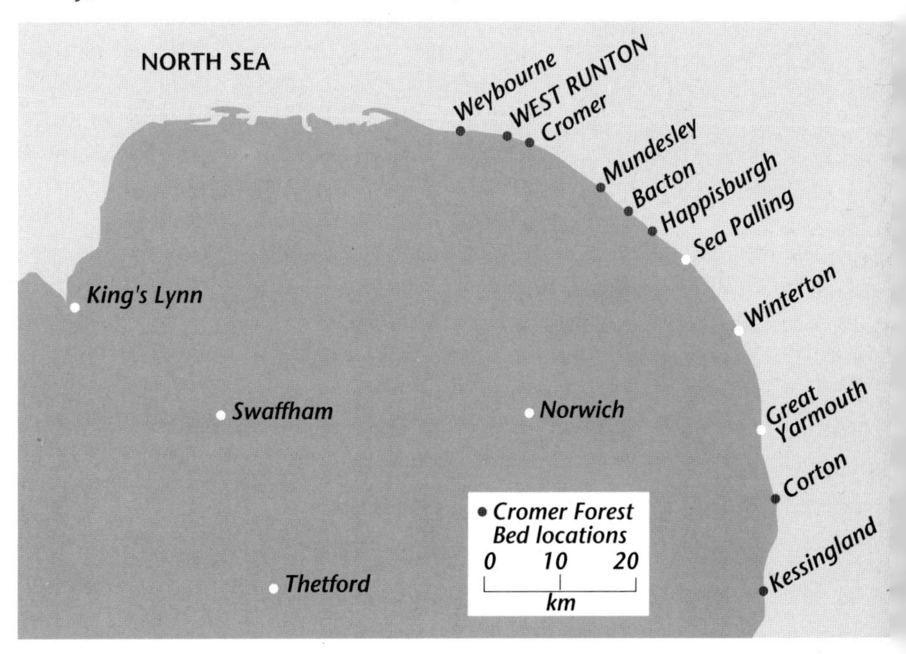

Since the early years of the nineteenth century the Forest Bed has been famous amongst geologists for its wealth of fossil mammal remains. Isolated bones and teeth, jawbones and the antlers of deer are found from time to time when these fossils have been eroded out of the cliffs or from deposits uncovered when storms occasionally wash away the beach. The Forest Bed and its fossils had an especially powerful attraction for Victorian geologists and fossil collectors. In the first half of the nineteenth century Anna Gurney, who lived at Northrepps in North Norfolk, obtained many important Forest Bed fossils by purchasing them from local fisherman. She generously donated this collection to the then Norwich Museum.

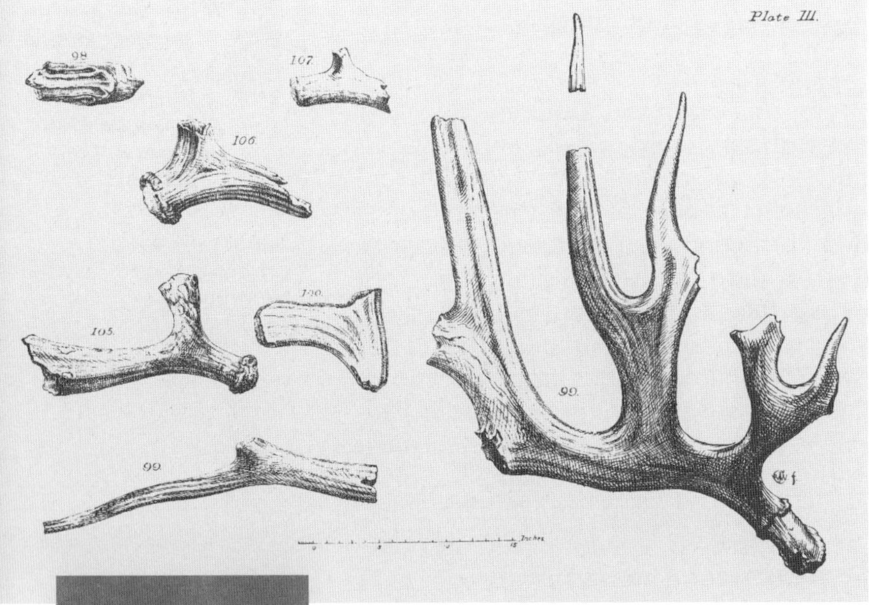

Antlers of extinct deer from the Forest Bed collected by John Gunn.

The Rev. John Gunn (1801-1890).

Another outstanding collector, and colourful character, was the Reverend John Gunn, a Norfolk vicar in the mid nineteenth century, who amassed a substantial collection of Forest Bed fossils, also now in Norwich Castle Museum. Alfred Savin, who kept a shop in Cromer, actively collected from the 1880's until the 1940's, and most of his specimens were purchased by the Natural History Museum in London.

Although today many of the classic Forest Bed sites are obscured by coastal defences, many important fossils finds continue to be found by amateur enthusiasts, many of whom work in conjunction with the professionals. There is a renewed interest in research on the sediments and fossils of the Forest Bed, which preserves a unique record of the changing climate and wildlife of a major part of the Quaternary period.

The West Runton Freshwater Bed at the base of the cliff.

The West Runton Freshwater Bed

The best place to see the Cromer Forest Bed is on the coast near the village of West Runton, two miles west of Cromer. Between about 200 and 300 metres east of the concrete slipway at West Runton Gap (also known as Woman Hythe) a prominent dark band can be seen at the base of the cliff. This dark band comprises organic-rich sediments filling an ancient river channel - the West Runton Freshwater Bed. The Freshwater Bed teems with fossils, ranging from microscopic pollen, seeds, mollusc shells and beetles to fishes, frogs, birds and large mammals. By comparing the kinds of fossils present with those from sites in the Netherlands and Germany, and by matching the land record with the well-dated deep ocean sequence, the age of the Freshwater Bed can be estimated at between 600,000 and 700,000 years.

The many fossils can tell us a great deal about the ancient environment of the area during the Cromerian period (named after the nearby town of Cromer), a warm or interglacial phase, which lasted for several thousand years. The presence of many species of plants and animals which occur in East Anglia today, and the absence of species now found further south in Europe, indicates that the climate was similar to that of the present day. Forest with oak, elm, pine, hazel and other trees clothed much of the landscape, with areas of wetland along the river valleys. The forests included areas of grasses and wild flowers, probably initiated and maintained by the activities of elephants and other large herbivores.

The wildlife was much more varied than anywhere in Europe today, and included such animals as hyaenas and bears, macaque monkey, wild boar, horses, bison, giant moose, rhinoceros -

and of course, elephant. However, at this time there were no people in the region, or indeed anywhere in northern Europe: as far as we know the first humans did not arrive in Britain until around 100,000 years later.

West Runton has been designated a Site of Special Scientific Interest by English Nature, and is the international type (standard) for the Cromerian warm period. The exceptional abundance and variety of the fossils from the Freshwater Bed make this a world class site.

Discovery of the Elephant

The story of the West Runton Elephant begins on 13 December 1990, when following a storm, naturalists Harold and Margaret Hems took a walk on the beach to see if any fossils had been uncovered by the sea. They found a large bone partly exposed in the face of the Freshwater Bed, and realizing its importance contacted Norfolk Museums Service. Excavation soon revealed that their find was the pelvic bone of a huge elephant. Next to the pelvis was a much smaller bone, an astragalus (ankle bone), also of an elephant. The occurrence of more than one elephant bone, apparently from the same individual, suggested that more bones might eventually come to light as the cliff was eroded by the sea.

The relatively narrow opening of the pelvic bone indicated that this was a male elephant, but there was no way of determining which kind of elephant it was. Three kinds of extinct elephant are known from teeth and bones found in the Cromer Forest Bed Formation, two species of mammoth *(Mammuthus meridionalis, Mammuthus trogontherii)*, and another species known as the straight-tusked elephant *(Palaeoloxodon antiquus)*. The West Runton pelvic bone and astragalus might have belonged to any of these. To make an identification, more bones were needed.

The 1992 Rescue Excavation

Fortunately, just over a year after the discovery of the pelvic bone, at Christmas 1991, the Freshwater Bed was once again eroded by heavy seas. This time several large bones exposed in the cliff were spotted by another keen eyed observer, Rob Sinclair, who also realised their importance and informed Norfolk Museums Service. It was now obvious that this was a find of major significance, and in January 1992 a rescue excavation was carried out by staff from Cromer Museum, Norwich Castle Museum, and volunteers.

This excavation recovered all the bones that could be reached without unsafe tunnelling into the cliff face - in total about a quarter of the skeleton, including most of the backbone, parts of the right front limb and, of particular importance for identifying the type of elephant, the lower jaw. One large limb bone, the right humerus, could not be excavated as it

Top left: An impression of the West Runton Elephant site 600,000 years ago; a scene in the Norfolk Broads today.

Top right: the pelvic bone under excavation.

Above: Harold and Margaret Hems.

Above: Rob Sinclair.

Left: Discovery of the pelvic bone.

went straight into the cliff. A convenient crack made it possible to remove part of the end of the bone, but the rest had to be left, protected by a sturdy small flint and cement wall built over the broken end.

Over a period of two years the new finds were painstakingly conserved by Rob Driscoll of the Castle Museum and volunteers Sonia Smith and Julie Curl. The lower jaw, now cleaned of adhering sediment, was the key to identifying the fossil remains. From the shortness of the jaw and relatively large number of enamel plates in the teeth, Dr Adrian Lister (University College, London) was able to identify the elephant as an early form of mammoth, *Mammuthus trogontherii*.

In common with other kinds of elephants mammoths originated in Africa. The first mammoths to reach Europe, about three million years ago, belonged to the species known as *Mammuthus meridionalis*. This evolved into *Mammuthus trogontherii* (the species found at West Runton), which in turn gave rise to the well-known but much smaller woolly mammoth *Mammuthus primigenius*, especially characteristic of the cold environments of the later part of the Ice Age. Woolly mammoths became extinct in Europe as recently as 12,000 years ago, but survived even later, until at least 4,000 years ago on Wrangel Island off northeast Siberia. Mammoths were a separate evolutionary branch of the elephant family which has completely died out. They were not the ancestors of the living African or Asian elephants.

From the bones recovered in 1990-1992 the height at the shoulder of the West Runton elephant when it was alive could be estimated at four metres, its weight at about ten tonnes - nearly twice the weight of a modern African elephant. This was a huge beast, amongst land animals exceeded in size and weight only by the very largest dinosaurs which had become extinct many millions of years earlier. From the state of wear of the teeth and degree of fusion of the bones the age at death can be estimated at forty years or so. Since an elephant's life span is similar to a human's, the West Runton elephant had died long before old age.

Elephant bones were not the only items found. There were also coprolites

(fossilized droppings) of spotted hyaenas *(Crocuta crocuta)*. Today these animals are restricted to Africa, south of the Sahara. Hunting in packs, they kill prey up to the size of zebra, and also scavenge carrion, crunching bones with their powerful jaws and teeth. Some of the West Runton elephant bones showed obvious signs of chewing, consistent with hyaena damage. Hyaenas had fed on the carcass as it lay in the river.

***Facing page**, from top: The 1992 rescue excavation;*
Ulna (diagonal) and other bones;
Lower jaw (centre) and other bones;
The lower jaw as found.

***This page:**, above left: Artist's impression of the West Runton elephant.*

Above, from top: Hyaena coprolite (fossil dropping) next to vertebrae;
Close-up of coprolite. Length 11cm;
Foot bone of the West Runton elephant. A large piece has been chewed off by a hyaena. Note tooth marks (arrowed). Length 11cm;
Modern spotted hyaena in Tanzania.

Above: Aerial view of 1995 excavation.

The 1995 Excavation

A quarter of the skeleton had been recovered. The rather tight scatter of the bones found in 1991-92 suggested that much more of the skeleton was present. But how was it to be retrieved from under twenty metres of cliff? Clearly a project to excavate the skeleton and record all the associated scientific information would need a great deal of planning, substantial finance, and the collaboration and goodwill of many individuals and organizations. Until the site could be excavated the bones were at considerable risk of destruction by the sea further eroding the cliffs during the winter months. This problem was solved by putting in place small-scale sea defences in the autumn of 1993.

Eventually major funding for the excavation was secured from the Heritage Lottery Fund (the first project on fossils to be funded by the Lottery), and from Anglian Water. The necessary permissions were obtained from English Nature and from the landowner David Abbs. A team of experts from several universities in Britain and the USA was assembled to research the sediments and fossil plants and animals from the site. The excavation was scheduled for the autumn of 1995. North Norfolk District Council Technical Services kindly offered to manage the engineering aspects of the project. Norfolk Archaeological Unit were asked to carry out the actual excavation. This was an unusual way to proceed, as archaeology deals with the works of humans, and this was an entirely natural site. However, it was felt that the skills of archaeologists in methodical excavation and recording would be invaluable in recovering this unique fossil.

Before the excavation could begin it was necessary to remove thousands of tonnes of overlying sands, silts and gravels using a dragline sited on top of the cliff to expose a surface of Freshwater Bed fifteen metres by five. The cliff had to be cut back to an angle of forty five degrees to provide a safe working environment for the team of archaeologists. On October 9th 1995 the small team of five people from the Norfolk Archaeological Unit, led by project manager Trevor Ashwin and deputy Sarah Bates, started to mark out the site and begin excavating.

As the beach material and other debris obscuring the site was cleared away, the three small cavities in the cliff face that had been dug in 1992 were rediscovered, remarkably intact after three and a half years, protected very effectively by the temporary sea defences. And sure enough there was the little flint and cement wall built over the part of the right humerus which was still in the cliff. This was very encouraging because it provided a direct link with the 1992 excavation and showed that the team was digging in the right place.

The project was boosted by substantial help from the Swedish consultancy Arkeologikonsult who provided state-of-the-art surveying and computer processing of site data. Totte Fors and Magnus Artursson from Arkeologikonsult surveyed the site and the cliffs using a total station theodolite (TST).

Below: Early stages of the excavation.

Bottom left: Surveying with the total station theodolite.

Bottom right: Removing the flint and cement wall from the right humerus (upper forelimb bone).

At a very early stage in the excavation a piece of elephant tusk turned up very near to the top of the Freshwater Bed. Associated with it was an elephant hyoid (tongue) bone - a very rare find indeed. However these remains were clearly not part of the main elephant skeleton as they were situated too high up. They were from another animal, or animals, which had lived and died some years later than the elephant that the team was looking for.

After two weeks of gradually excavating down from the top, at last elephant bones began appear, much to the relief of the Project Director. Logic said that much more of the skeleton should be there, but there was always a nagging doubt that after going to all this trouble and expense perhaps nothing very much would be found after all!

Shortly after discovering the left ulna (lower forelimb bone), work began on uncovering the right humerus (upper forelimb bone), which had been left in the cliff in the 1992 rescue excavation. The flint and cement wall encasing the broken end had to be carefully chipped away before the bone could be excavated further. The safe retrieval of the elephant bones was the responsibility of Gordon Turner-Walker and Elena Makridou of the Conservation Department of Norfolk Museums Service. They used the old and well-tried technique of plaster-bandaging, very much as used to support and protect a broken leg. The plaster jacket solves the problem of moving a large and heavy object that has lost its inherent strength and so could break up when moved. Firstly the bone is partially excavated, leaving it on a pedestal of sediment. Then layers of tissue and aluminium foil are applied to

Above: Both femurs (thigh bones) as found.

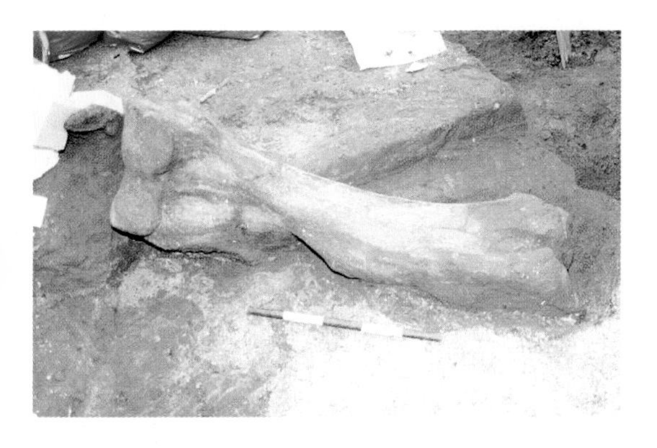

Left, from top: Left humerus (upper fore-limb bone); Excavation of two thoracic vertebrae; Left scapula (shoulder blade) overlying neck vertebrae; Plaster-bandaging the left tusk.

Below: Excavating the plaster-bandaged right femur (thigh bone).

prevent plaster sticking to the bone surface, followed by layers of bandage (first fine, then coarse) dipped in wet plaster which rapidly set into a hard coat. For larger bones the plaster jacket is reinforced with metal mesh, and for the biggest bones lengths of wood or angle iron are built in. When the jacket has hardened, the remaining sediment is cut away and the bone lifted and carried off the site.

More and more bones appeared, including nearly all of the large limb bones, more vertebrae and the prize find of the skull. With the earlier finds, we now had a near-complete skeleton, by far the best and most impressive fossil ever found in the Cromer Forest Bed Formation. As bones were uncovered, singly or in groups, each was carefully recorded prior to plaster-bandaging and lifting. As well as recovering the bones for future study and display, it is also very important to accurately record their distribution in the ground and their relationships to one another. As elephant bones were found their positions were recorded by TST, and in addition accurately plotted by hand using a planning frame. By means of a laser beam, the TST computes the distance very accurately to a staff positioned at various points on a bone, or any other object. The data are downloaded onto a computer and used to generate plans of the site. This information can tell us a great deal about what happened to the carcass of the elephant, and possibly even how it died.

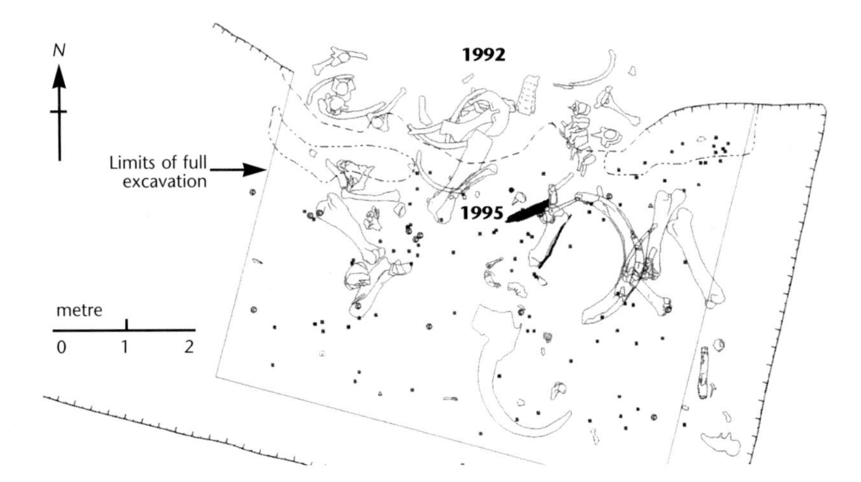

By the end of November, after seven weeks, the Norfolk Archaeological Unit had completed their work on the site, and along with Arkeologikonsult they withdrew to their offices to begin the work of producing the final plans and records of the site.

Now only the skull remained. The weather was deteriorating, the site had already survived one exceptionally high tide, and the alarming possibility that the site could be flooded by the sea seemed more and more likely. Two weeks were all that were left for a small team of museum staff and volunteers to lift the skull to safety.

Dr Gordon Turner-Walker, the museum's archaeological conservator, and Graham Powell, from the Sainsbury Centre for Visual Arts in Norwich, solved the problem by building a steel cage on site to take the weight of the skull. Under the glare of media attention, the skull was successfully lifted using a mechanical excavator and transported off the beach.

***Opposite page**, top left: The skull from the front.*

Far left: Using a planning frame to plot the left tibia.

Bottom left: Plan of the elephant bones.

Right: Constructing a steel cage to support the skull.

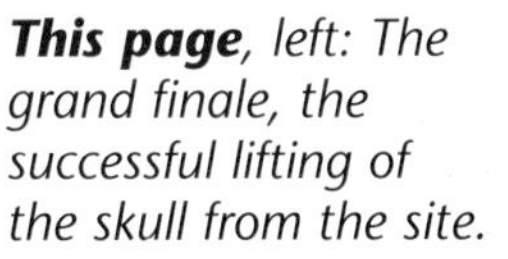

***This page**, left: The grand finale, the successful lifting of the skull from the site.*

Research

In the past the West Runton Freshwater Bed has been the subject of several research studies, but these were unco-ordinated and no work had been done on sediments, beetles and other important aspects of the site. The discovery and excavation of the elephant has focussed scientific attention on the site, bringing together a team of research workers, with the happy result that an integrated research programme is now

Above: Taking sediment samples for fossil pollen.

being carried out, including work on the sediments, chemistry and fossils.

Although seldom obvious when digging the deposit, tiny bones, mollusc shells and seeds can be retrieved by washing the sediment through sieves, and sorting the residue. The sediments themselves and the many other animal and plant fossils that they contain will provide a wealth of information on the environment and wildlife of what is now East Anglia more than half a million years ago.

Samples of up to a few kilogrammes are needed to be sieved in the laboratory for seeds and fruits, mollusc shells, beetles and other fossils. These fossils tell us mainly about the plants and animals living in the immediate vicinity of the site.

Still larger samples, over ten tonnes in all, were taken to be sieved for small-vertebrate bones and teeth. This will be a major operation, but will yield tens or hundreds of thousands of fossils, very probably including species previously unknown for this period.

Microscopic pollen grains are preserved in very large numbers in suitable sediments, such as the fine-grained sediments of rivers and lakes. Pollen is produced in large quantities by all higher plants, especially those pollinated by wind. The pollen grains can be identified under the microscope to discover which trees and other plants grew at various time in the past. Because pollen can be blown for some distance it gives us information about the overall vegetation cover of an area, not just the plants growing locally. Only very small samples of sediment are needed as each contains thousands of pollen grains. Sediment profiles through the Freshwater Bed were sampled at 2cm intervals - the series of samples will reveal changes in the vegetation over the period of time that it took to deposit the bed.

Fine grained sediments can preserve a record of the earth's magnetic field at the time that the bed was deposited and samples taken from the Freshwater Bed may give important information on these changes, and provide clues to dating the bed. The types and sequences of sediment were carefully recorded and will provide information on the ancient river. Samples were also taken to analyse the sulphur content - providing data on the

Right: The skull as found (top missing), with the right tusk in place.

Above: Evidence of trampling by other elephants; damaged left tusk, and end of right femur pushed into the sediment. (scale = 1 metre)

Left: Detail of damage to left tusk, before removal of tusk fragments. (scale = 0.5 metre)

salinity of the West Runton river. Other samples were taken for oxygen isotopes which can provide information on past temperatures.

Death and Burial

Until much more work on the remains and site has been completed, it is not possible to say why the elephant died, although it clearly died prematurely. Nevertheless, we can say quite a lot about what happened to the elephant after it died. A glance at the plan shows that the bones were disarticulated and jumbled before burial, but closer inspection reveals that there is a pattern to the distribution - although several bones are noticeably out of place, generally speaking the front end of the animal was to the west and the back end towards the east. However, the skull was found at the wrong end!

Several of the larger bones were found resting at steep angles. The left tusk was generally in good condition, but flattened and damaged. The middle section was strangely crushed and broken. Beneath the broken part of the tusk one end of the right femur appeared to have been forced down into the sediment. Work on the bones is still at an early stage, but it seems likely that other elephants visited the site perhaps a few years after the death of our elephant and moved some of the bones. This behaviour is commonly seen in elephants today, who are strangely attracted to the remains of their own species. In the process they probably trampled many of the bones into the soft sediment.

The 1995 excavation provided more evidence of hyaena activity, including several more coprolites. Some of the smaller elephant bones including most of the foot bones are missing and may have been removed or eaten by hyaenas, but they would have been quite unable to move the large limb bones or jaw.

The likely reason for the odd position of the skull is perhaps surprising. Modern elephant skulls have been observed to float in water, buoyed up by the numerous air-filled cavities which they contain. The skull of the West Runton elephant may have broken loose from the rotting carcass and drifted a few metres with the current. The striking upright position in which it was found is also consistent with the skull having acted as a float. The relatively fragile top of the skull, which contained the air spaces, was missing, probably because it had disintegrated within a few years of death due to exposure to the elements.

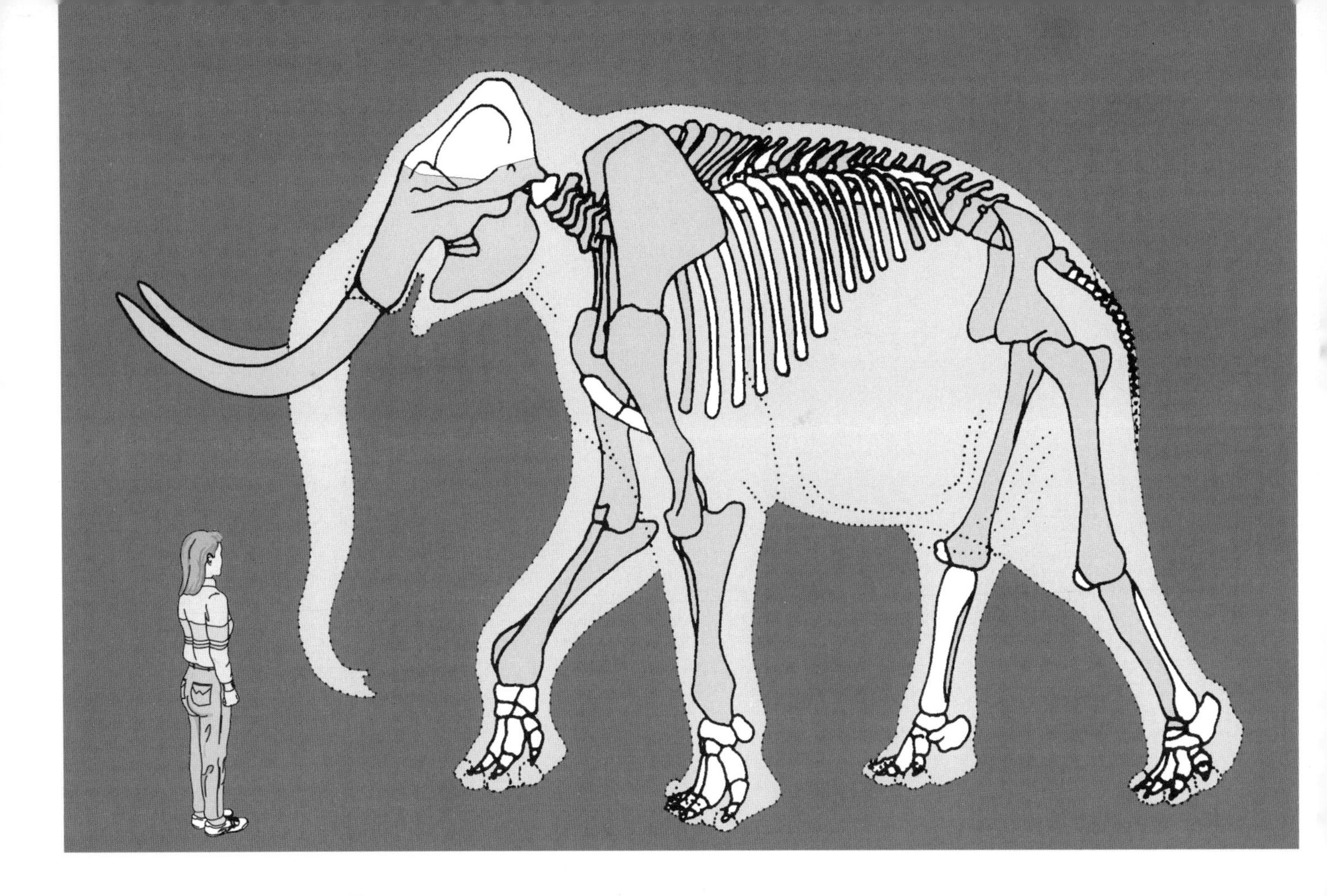

Above: Outline of the West Runton Elephant showing bones found 1990-95. Identification of specific ribs is uncertain at present.

Right: Conservation work on the left femur (thigh bone).

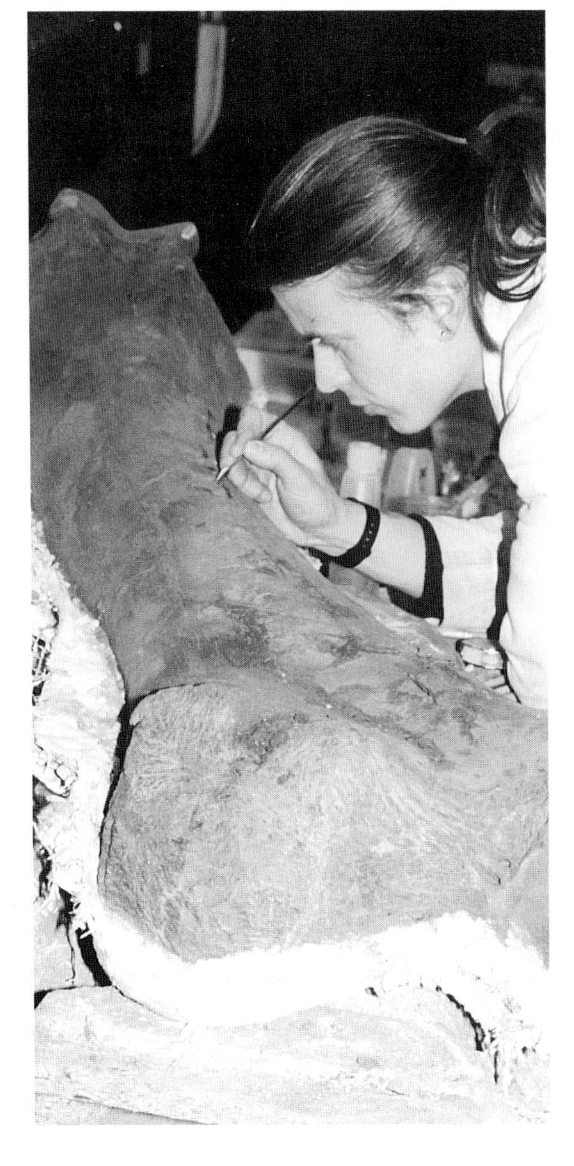

What happens now?

Most of the skeleton of this huge prehistoric elephant had now been rescued, together with a wealth of information about its ancient environment and how it had lived and died.

It took nine weeks to excavate the elephant, but the careful task of conserving the bones is likely to take a few years to complete. Each bone has to be gradually unwrapped from its plaster jacket, cleaned of sediment and where appropriate treated with plastic consolidants. Fibreglass supports will be needed for the larger bones which cannot support their own weight.

As the post-excavation work proceeds, on the one hand on the conservation and study of the elephant remains, and on the other on the Freshwater Bed and its abundant fossils, much new, and probably unexpected, information will come to light. All this work promises to give an extraordinarily full picture of the environment and wildlife of the region more than half a million years ago.

The bones, a full sized replica of the skeleton and information about the ancient site will be put together in a permanent display in Norwich Castle Museum.

Photographs
Derek A. Edwards:
p.6 aerial views;
Harold A. Hems:
p.3 discovery, p.5 hyaena;
Alan Howard:
p.5 coprolite, chewed bone;
Elena Makridov:
p.11 detail of damage;
Neil Moss: p.7 skull,
p.11 evidence of trampling;
Martin Warren: p.6 surveying,
p.8 constructing skull
support (bottom),p.9;
David Wicks: cover,
p.10 skull as found;
Tony Stuart: inside covers,
p.2 West Runton cliff,
p.3 Norfolk Broads, pelvic
bone, p.4, p.5 hyaena
coprolite as found,
p.6 early excavation,
removing cement wall,
p.7, p.8 constructing skull
support (top),
p.10 sampling for pollen,
p.12 conservation of femur.

Illustrations
Sam Brown:
Front cover; p.5 West
Runton Elephant;
Arkeologikonsult:
p.8 site plan;
Ashley Sampson
p.12 Elephant outline.

Further Reading

The West Runton Elephant, Norfolk Museums Service poster, 1993.

Ashwin T and Stuart A J, 1996, 'The West Runton Elephant', *Current Archaeology 149*, September, 164-168.

Jones R L and Keen D H, 1993, *Pleistocene Environments in the British Isles*, London: Chapman and Hall.

Lister A M and Bahn P, 1995, *Mammoths*, London: MacMillan.

Stuart A J, 1982, *Pleistocene Vertebrates in the British Isles*, London: Longman.

Stuart A J, 1988, *Life in the Ice Age*, Princes Risborough: Shire Publications.

Stuart A J, 1989, *The Ice Age In East Anglia*, Norfolk Museums Service Information Sheet.

Sutcliffe A J, 1986, *On The Track of Ice Age Mammals*, London: British Museum (Natural History).

ISBN 0-903101-64-5
9 780903 101646 >